BIG AND SMALL

Julia Wall

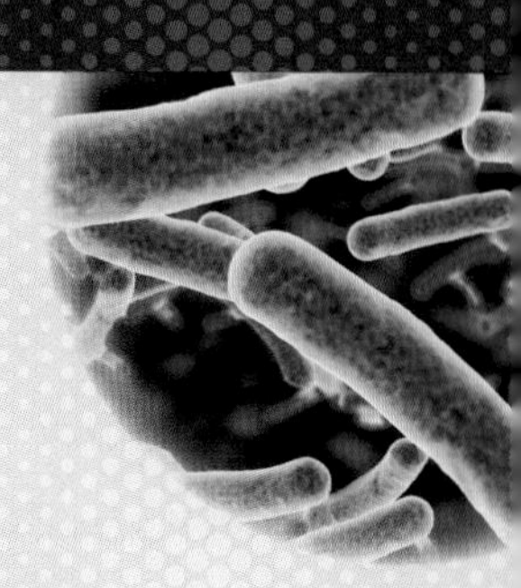

CONTENTS

HOW BIG ARE YOU?

Have you ever wondered how big you must look to an ant? What about how small you might look to a blue whale?

People have always been curious about things that are very small and very big. Scientists study **objects** and **structures** that are extremely small or big. With the help of technology, they measure things as tiny as atoms and as large as galaxies.

FOCUSING LIGHT

A lens in a telescope

Around two thousand years ago, people were learning how to make glass from different materials. At first, they used glass to make bottles and jars, but soon they were making windows, mirrors, and many other things. About a thousand years ago, people discovered how to grind glass to make a lens. Lenses are pieces of curved glass that can be used to make things look bigger and closer. By the 1600s, scientists had worked out how to put several lenses into a tube to make a microscope or a telescope.

Did you know that when lenses were first invented, people thought they looked like lentil beans? That's why they're called lenses!

Microscopes are one way technology helps us to study things that are very small. They enable you see extremely small things. Telescopes allow you see big things that are very far away. They both work by **focusing** light.

Lentils

MICROSCOPES

When they were invented, microscopes let people look at common objects in a new way. Scientists looked through microscopes and saw objects **magnified** many times—they made small things look bigger.

A snowflake, as seen through a microscope

They also used microscopes to study things that no one had ever seen before. When scientists looked through microscopes, they discovered tiny things like bacteria. At first, they didn't know what these tiny things were!

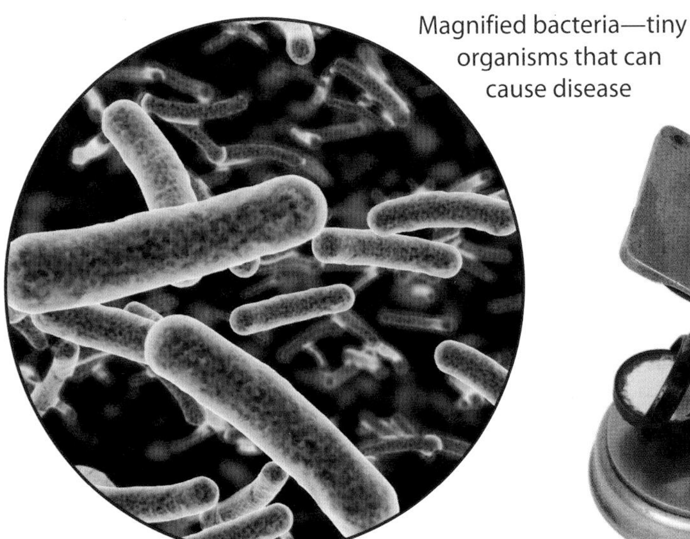

Magnified bacteria—tiny organisms that can cause disease

One of the first microscopes—invented about 400 years ago

If you were to look through a microscope, you would see some of the objects that amazed scientists long ago. Try looking at one of your hairs through a microscope.

A human hair magnified ten times

Electron Microscopes

Using an ordinary microscope, we can now magnify things about 2,000 times. **Electron** microscopes make it possible to magnify things even further. Instead of focusing light on an object, electron microscopes focus a beam of electrons. An interaction between the beam and the object is then projected onto a fluorescent screen, a bit like the one in your TV, creating an image.

A transmission electron microscope

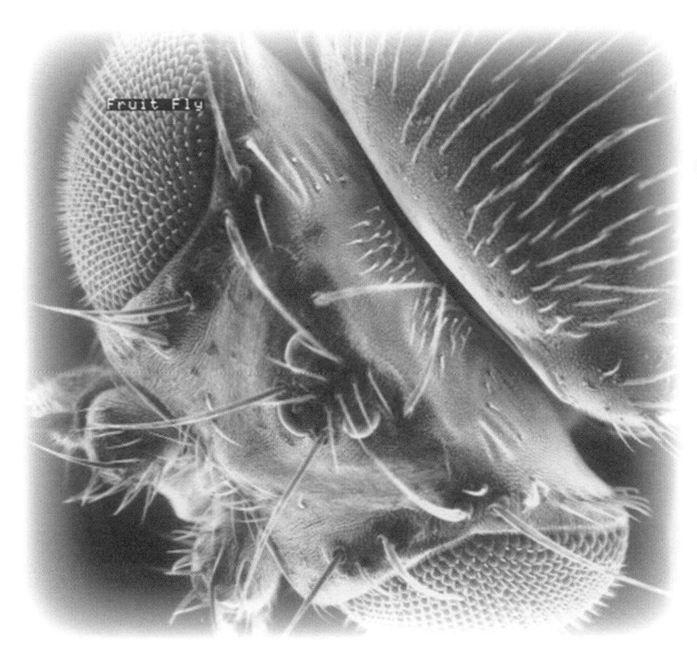

Did you know that an electron microscope can magnify an object more than a million times?

A fly's face magnified 110 times by an electron microscope

TELESCOPES

Have you ever looked through a telescope? By collecting and magnifying small amounts of light, optical telescopes allow us to see the details of objects such as stars, planets, and moons, even though they are very far away.

The moon, as seen with the naked eye

The moon, as seen through a telescope

The power of a telescope depends on the size of its lenses or mirrors. Bigger lenses and mirrors collect more light than smaller ones. Bigger lenses make faraway objects easier to see.

Many optical telescopes use mirrors, instead of lenses, to collect light.

Galileo Galilei lived in Italy from 1564 to 1642. He was a scientist who was interested in light, **gravity**, and the movement of objects. In 1610, he pointed a telescope toward the planet Jupiter and discovered four large moons.

Did you know that Jupiter actually has more than 60 moons? Its four biggest moons, named Io, Europa, Ganymede, and Callisto, are called "the Galilean moons."

Galileo was one of the first scientists to realize that the earth and the other planets in our solar system **orbit** the sun. During Galileo's time, many people didn't believe this. They thought the sun and other planets circled the earth.

The planets orbit the sun.

This diagram is not to **scale** because this page isn't big enough to show such huge distances.

Galileo's telescope

People called Galileo's telescope "an instrument of the devil" because they thought it was telling a lie.

The Hubble Telescope takes photographs of space from space. It is the size of a school bus and orbits the earth every 97 minutes. The photographs it takes are clearer than those taken from earth because earth's **atmosphere** isn't in the way.

Hubble has a giant mirror that scientists can point in different directions. A computer turns the light Hubble captures into pictures. These are relayed from Hubble to a satellite and then to computers here on earth.

Saturn, as seen by the Hubble Space Telescope—Saturn is never less than 746 million miles (1.2 billion kilometers) away from the earth

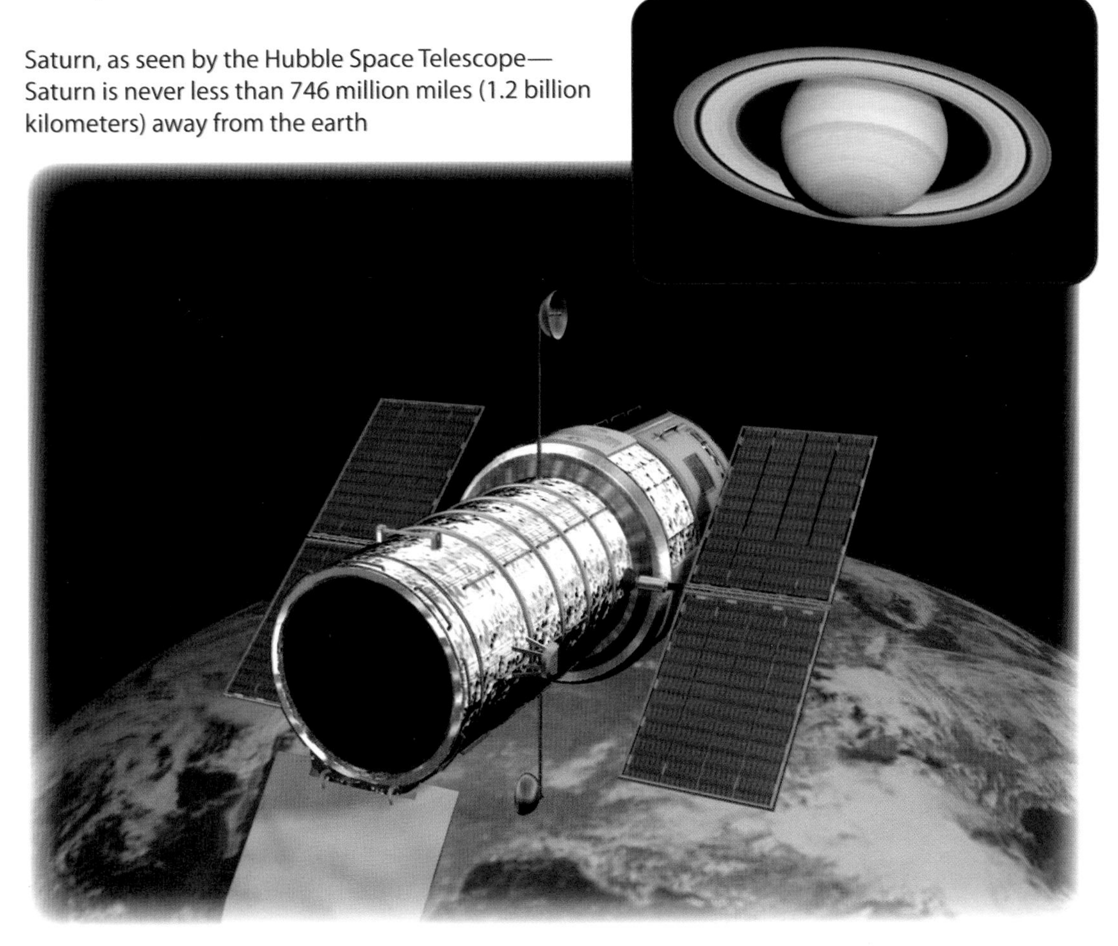

The Hubble Space Telescope

Looking for Aliens

The technology behind telescopes helps us in other ways, too. A different kind of telescope—a radio telescope—gathers radio signals from distant stars and huge clouds of hydrogen gas drifting in space. These signals don't come from radio broadcasts, though! They arrive in the form of **electromagnetic waves**.

Our sun is a star. It sends out radio signals.

SETI (the **s**earch for **e**xtra**t**errestrial **i**ntelligence) looks for evidence of **space aliens**. SETI collects information using radio telescopes.

SETI isn't just one radio telescope—it's lots and lots of computers and radio telescopes working together in a huge network.

MEASURING BIG AND SMALL OBJECTS

Thanks to technology, we have lots of ways to measure large and small objects.

An artist's impression of the sun seen from Quaoar

Measuring Objects in Space

You've probably stood back-to-back with a friend or family member to see who's taller. Scientists sometimes measure objects in space by comparing the objects to things that are behind them.

When scientists found a new planet-like object in 2002, they named it Quaoar (pronounced *Kwaa-waar*). Scientists **estimate** the size of Quaoar from the amount of light it reflects and from the size of the objects behind it. From this, they estimate that Quaoar has a **diameter** of about 745 miles (about 1,199 kilometers). By comparison, Texas is 790 miles (1,271.38 kilometers) from north to south.

Quaoar is even further away from the earth than Pluto is! Its orbit is 3.7 billion miles (6 billion kilometers) from the sun.

Measuring Whales

Even if you had a measuring tape that was long enough, how could you get a whale to hold still while you measured it? Living whales are hard to measure.

A whale's blowhole

But technology gives us a way. Scientists measure the **blowholes** of dead whales that wash up on beaches. This information tells them how big the blowholes of different sized whales are. They've discovered that the larger the whale, the longer it takes for a sound to come out of its blowhole.

Living whales make sounds through their blowholes. By measuring how much time passes between each sound, a scientist can estimate the size of a whale's blowhole. Once you know the size of the blowhole, you can estimate the size of the whale!

Dinosaur Math

Dinosaurs have been **extinct** for about 65 million years, but scientists are still discovering many things about them. Just like detectives sometimes use footprints to help solve crimes, scientists use dinosaur footprints to answer questions about dinosaurs. They have figured out that the nose-to-tail length of a dinosaur equals 10 times the length of its footprint.

From these footprints, we can tell the length of the dinosaur that made them.

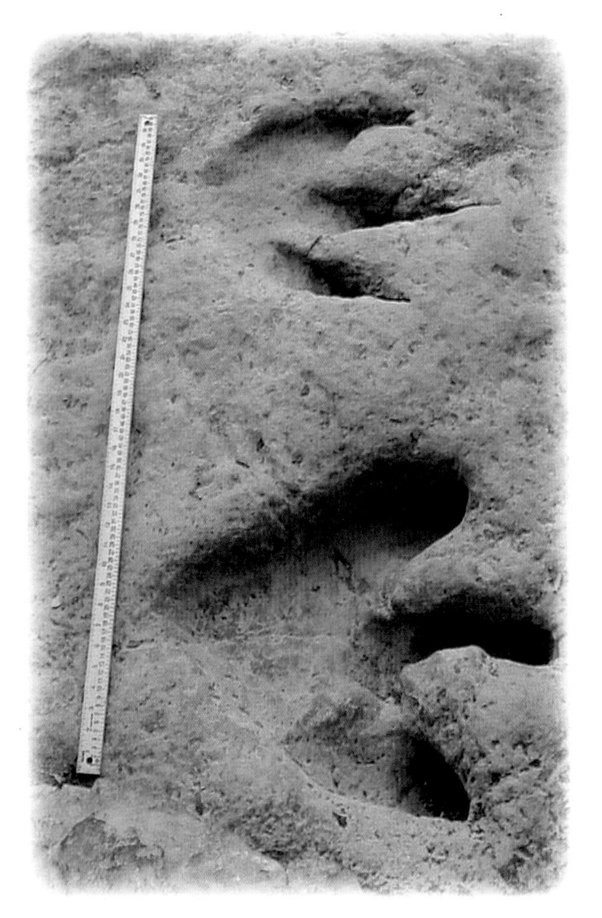

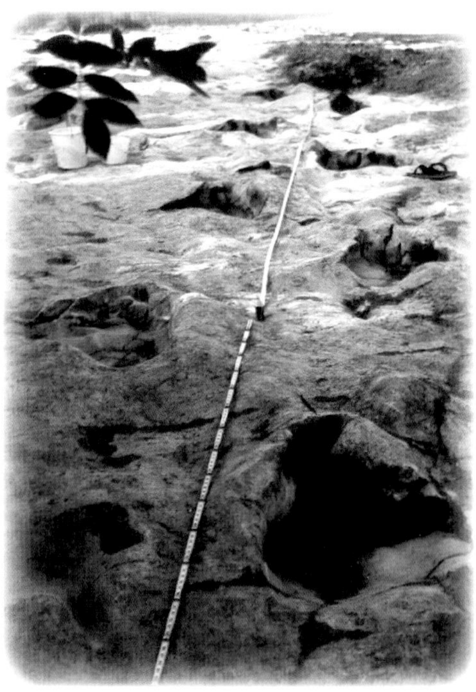

The largest footprints found so far belong to plant-eating dinosaurs called sauropods. These sauropod tracks were found in Texas. Each footprint is about 3.3 feet (1 meter) long, so the dinosaur that made them was 33 feet (10 meters) from nose to tail.

Find Your Dinosaur Length

If you were a dinosaur, how long would you be from head to toe?

You will need:

- chalk
- a ruler
- a pencil and paper

What to do:

1. Take your shoes and socks off.
2. Trace the outline of one of your feet with a piece of chalk.
3. Measure the length of your footprint and write it down.
4. Multiply this measurement by 10. The answer tells you how long you would be if you were a dinosaur!

Name	Length of footprint	x10	= dinosaur length
		x10	
		x10	
		x10	

Compare your result:

Get others to repeat the experiment with their feet. Explain to your friends how you know how long they would be if they were dinosaurs. Which of you would be the longest dinosaur?

Nanotechnology

We measure large objects using measurements such as yards and meters. For objects too small to see with our naked eye, such as atoms, we can measure them in nanometers. To get an idea of just how small this is, if you were to split a hair evenly into 50,000 **strands**, each strand would be one nanometer wide.

$$\textbf{nano} = \frac{1}{1{,}000{,}000{,}000}$$

Nanotechnologists first "saw" a single atom, using a spectroscope, in 2004.

A spectroscope breaks white light into the colors of the rainbow.

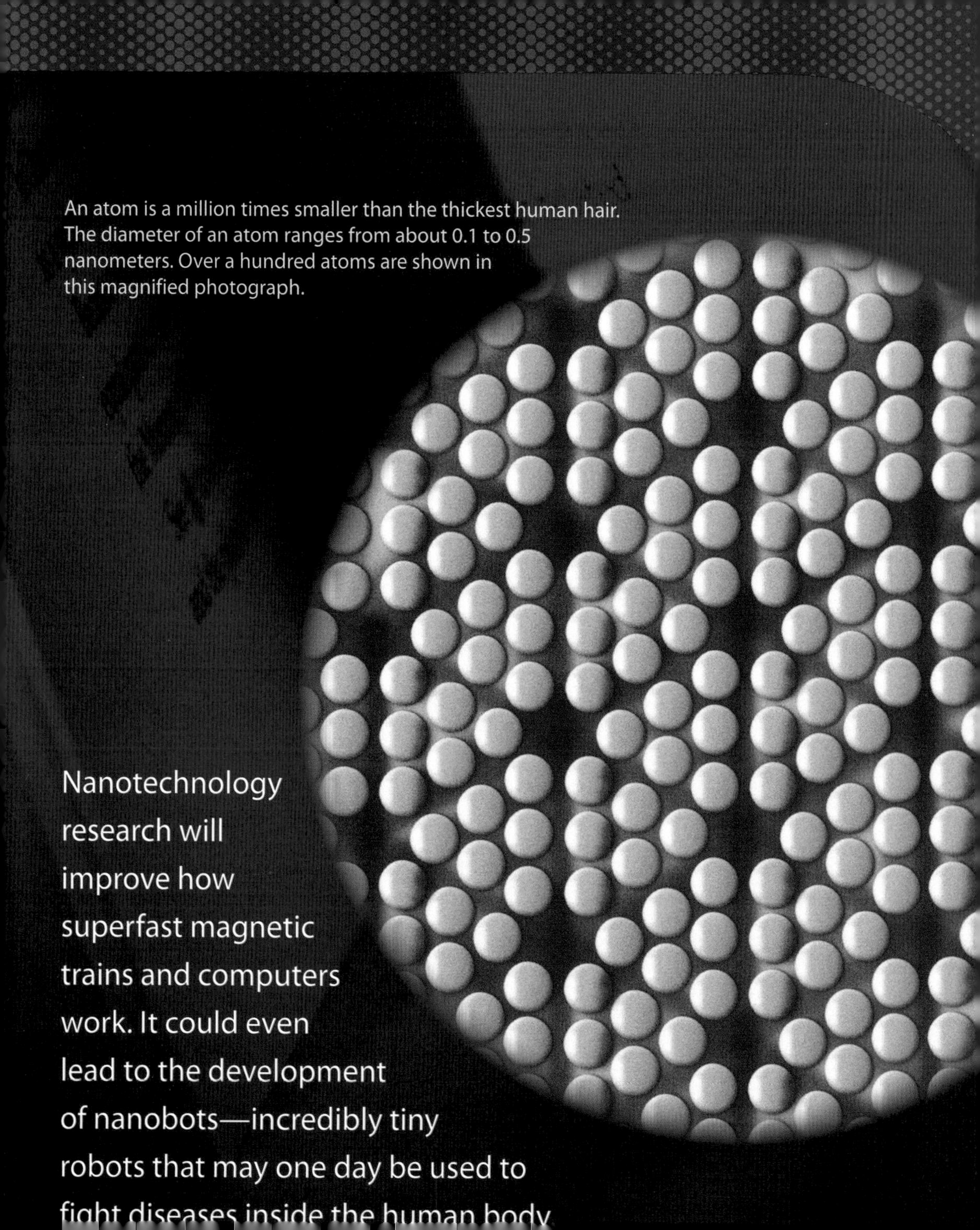

An atom is a million times smaller than the thickest human hair. The diameter of an atom ranges from about 0.1 to 0.5 nanometers. Over a hundred atoms are shown in this magnified photograph.

Nanotechnology research will improve how superfast magnetic trains and computers work. It could even lead to the development of nanobots—incredibly tiny robots that may one day be used to fight diseases inside the human body.

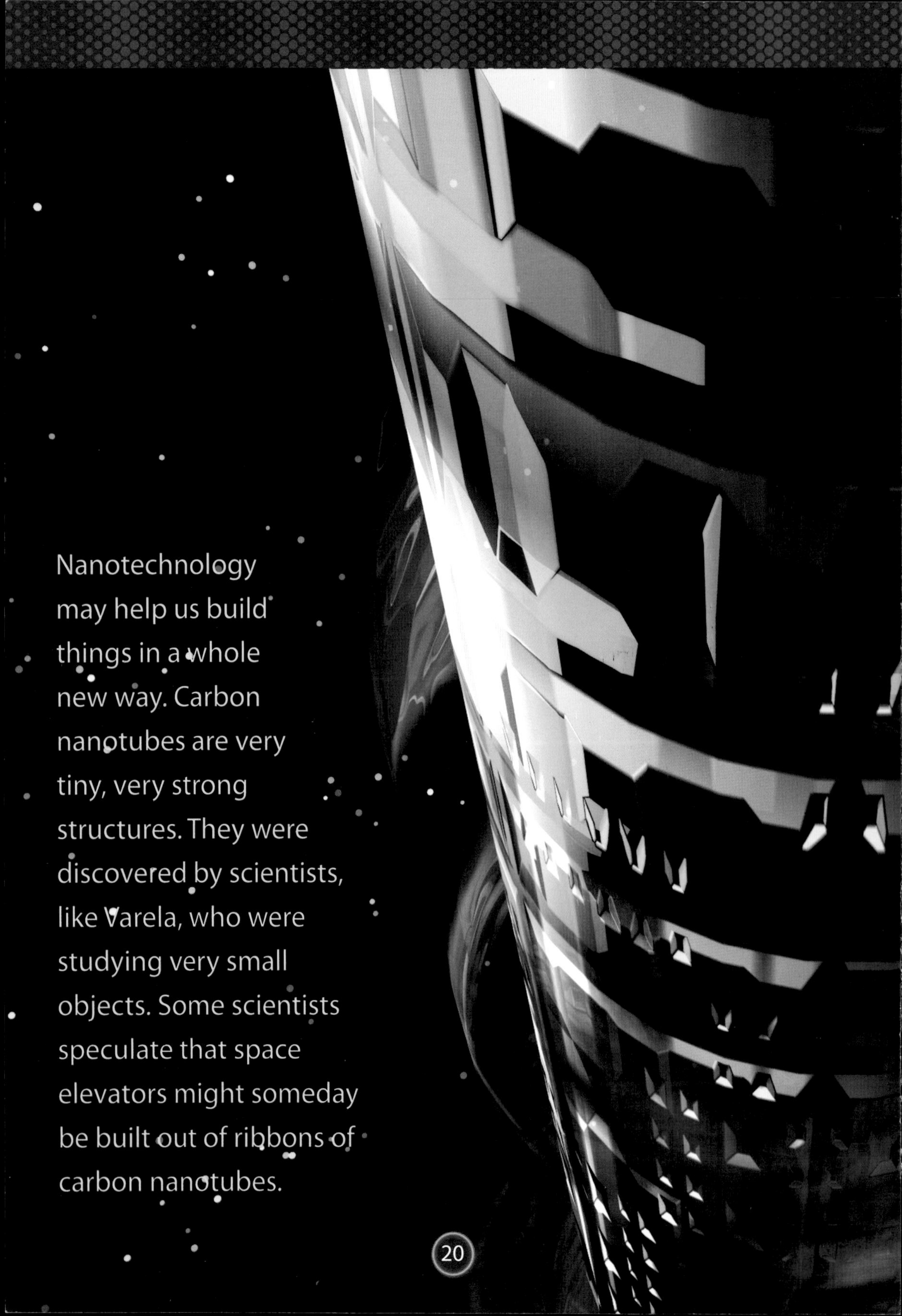

Nanotechnology may help us build things in a whole new way. Carbon nanotubes are very tiny, very strong structures. They were discovered by scientists, like Varela, who were studying very small objects. Some scientists speculate that space elevators might someday be built out of ribbons of carbon nanotubes.

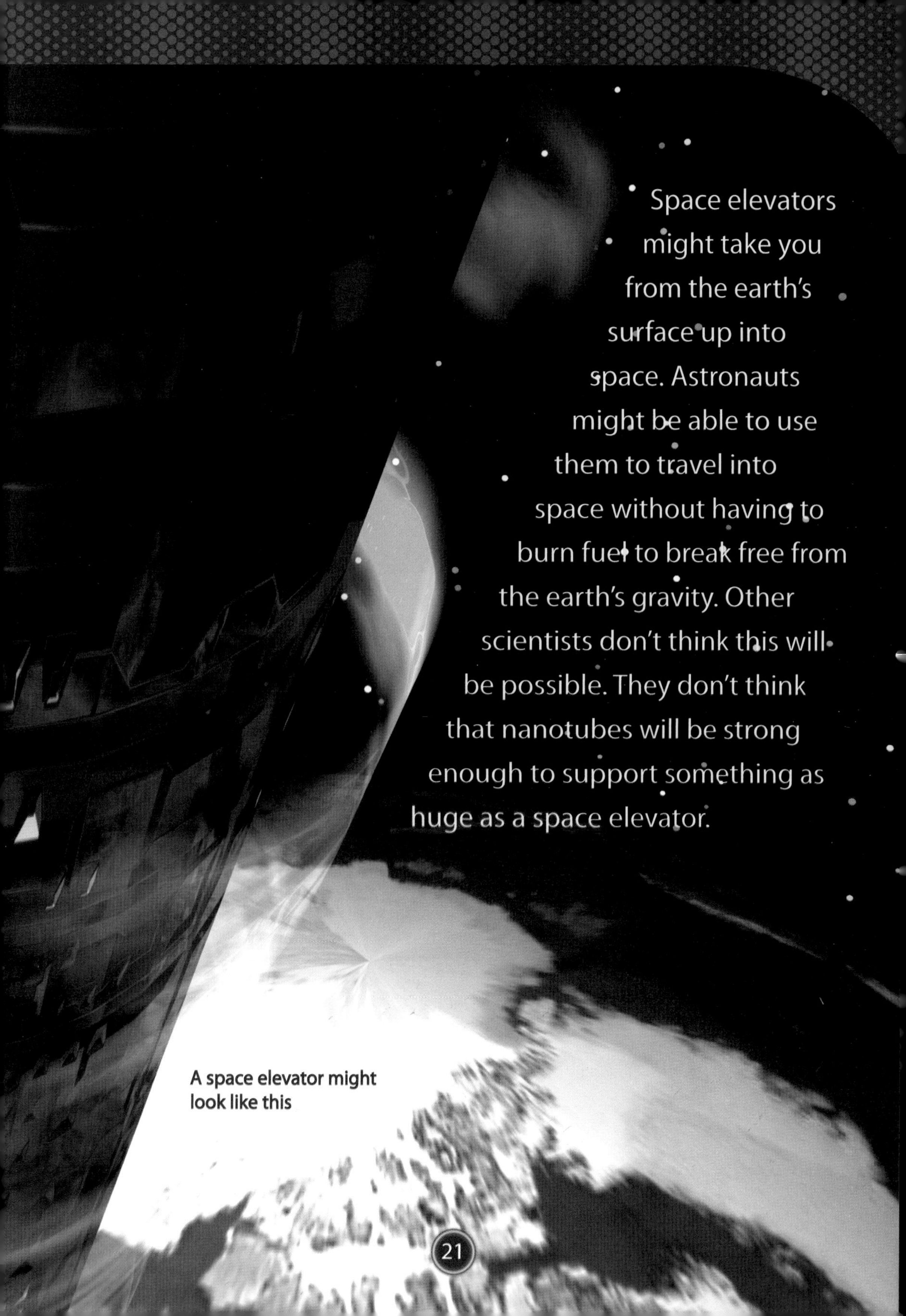

Space elevators might take you from the earth's surface up into space. Astronauts might be able to use them to travel into space without having to burn fuel to break free from the earth's gravity. Other scientists don't think this will be possible. They don't think that nanotubes will be strong enough to support something as huge as a space elevator.

A space elevator might look like this

COMMUNICATING WITH ALIENS

If the SETI project ever does locate aliens, how will aliens "see" and measure big and small things? Will their point of view be that of something small, such as an ant, or that of something big, such as a planet? In order to communicate with alien scientists, we will need a common way of measuring large and small things, such as the distance between earth and where they live.

A laser beam of light

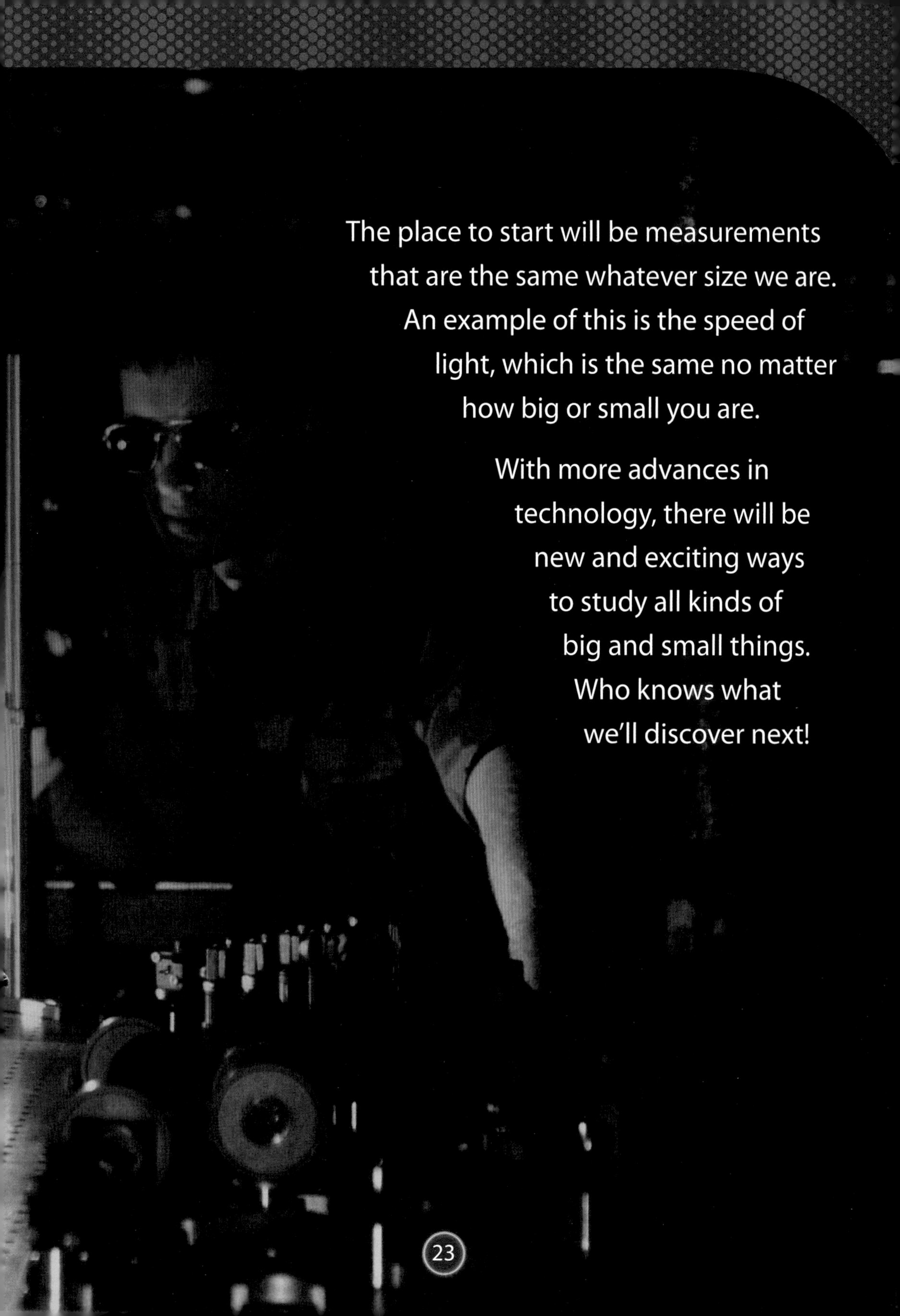

The place to start will be measurements that are the same whatever size we are. An example of this is the speed of light, which is the same no matter how big or small you are.

With more advances in technology, there will be new and exciting ways to study all kinds of big and small things. Who knows what we'll discover next!

GLOSSARY

atmosphere—the layer of gas around a planet

blowholes—nostril-like openings at the top of whales' heads

diameter—the distance across a sphere or circle

electromagnetic waves—electric and magnetic pulses

electron—an electrically charged part of an atom

estimate—make a rough calculation

extinct—no longer existing

focusing—using lenses to bring light rays to one spot

gravity—a force that an object exerts on another

magnified—made bigger

objects—things

orbit—travel around in a circular or elliptical path

scale—in proportion to the way it is in reality

space aliens—beings from outer space, extraterrestrials

strands—long, thin pieces

structures—an arrangement of objects of things